Plants on the Move

Émilie Vast

Translated by Julie Cormier

iɴi Charlesbridge

2021 First US edition
Copyright © 2018 éditions MeMo
First published as *Plantes Vagabondes* by éditions MeMo
English translation © 2021 by Charlesbridge

Published by Charlesbridge
9 Galen Street
Watertown, MA 02472
(617) 926-0329
www.charlesbridge.com

Type set in Meta by Erik Spiekermann and Rotis Semi Serif by Adobe
Color separations by Colourscan Print Co Pte Ltd, Singapore
Printed by 1010 Printing International Limited in Huizhou,
    Guangdong, China
Production supervision by Jennifer Most Delaney
Designed by Cathleen Schaad

LIBRARY OF CONGRESS CATALOGING-IN-PUBLICATION DATA
Names: Vast, Émilie, 1978-author.
Title: Plants on the move / Émilie Vast.
Description: Watertown, MA: Charlesbridge, [2021] | Audience: Ages 5-8. |
    Summary: "Examples of the many ways plants and their seeds move—
    falling, clinging, burrowing, floating, via animal consumption or
    transport—in their quest to reproduce"—Provided by publisher.
Identifiers: LCCN 2020000778 (print) | LCCN 2020000779 (ebook) | ISBN
    9781623541484 (hardcover) | ISBN 9781632899255 (ebook)
Subjects: LCSH: Plants—Dispersal—Juvenile literature. | Seeds—Dispersal—
    Juvenile literature. | Plant ecology—Juvenile literature. | Phytogeography—
    Juvenile literature.
Classification: LCC QK929 .V37 2021 (print) | LCC QK929 (ebook) | DDC
    581.4/67—dc23
LC record available at https://lccn.loc.gov/2020000778
LC ebook record available at https://lccn.loc.gov/2020000779

Printed in China
(hc) 10 9 8 7 6 5 4 3 2 1

I'm a dandelion.

This yellow disk that I turn toward the sky is my flower— or, if you look closely, my flowers. They are very tiny and nestled tightly together, waiting for insects to arrive.

Insects move pollen from flower to flower.

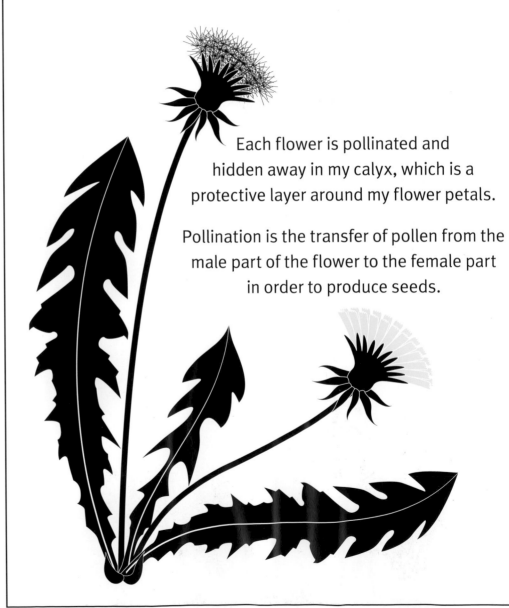

Each flower is pollinated and hidden away in my calyx, which is a protective layer around my flower petals.

Pollination is the transfer of pollen from the male part of the flower to the female part in order to produce seeds.

See the flowers transform. . . .

Each one becomes a seed,
an incredible flying seed!

And with a puff of your breath . . .

you send them into the breeze to pass through fields and gardens.

They will find somewhere to settle,

they will shed their parachutes,

sprout,

and then . . . grow!

Other flowers that fly away:

Horseweed

Milk Thistle

Ragwort

Creep

I'm a strawberry plant. You know me well, and you love to eat my sweet, red fruit. In order for you to have plenty of strawberries to enjoy, you need strawberry plants.

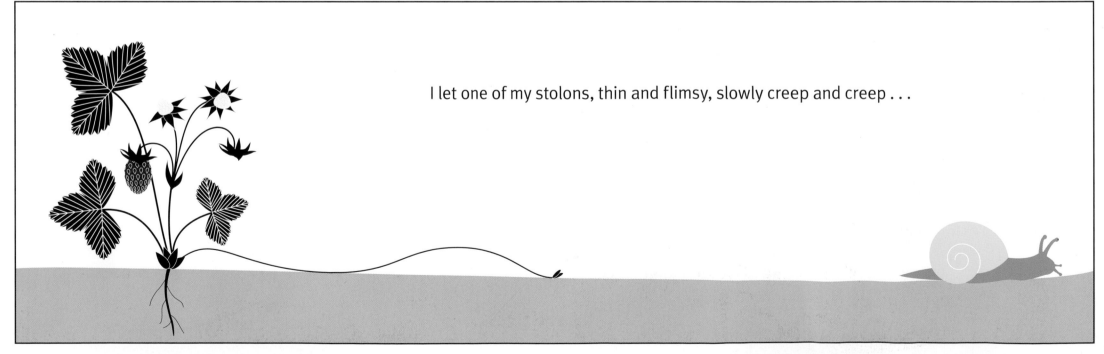

I let one of my stolons, thin and flimsy, slowly creep and creep . . .

until it finds a place it likes. Then it settles in and makes little roots. A mini-me is born!

This new little plant grows up
and makes flowers,
which then become fruits.
And then the cycle repeats. . . .

Other plants that creep:

Periwinkle

Cinquefoil

Buttercup

Fall
and Whirl
and
Bounce

We're chestnut and maple, trees
so big that our branches touch the sky.

The summer has ended, and our flowers
have transformed. One into beautiful,
prickly spheres, and the other into light,
delicate wings.

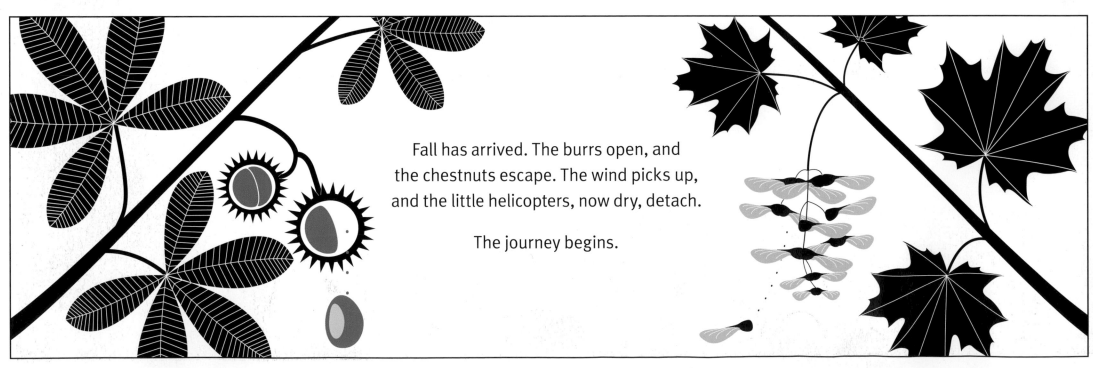

Fall has arrived. The burrs open, and
the chestnuts escape. The wind picks up,
and the little helicopters, now dry, detach.

The journey begins.

A chestnut falls and bounces
like a ball at the foot of the maple tree.
The maple key, or samara, spirals
down in the wind to the roots of the
chestnut tree.

Mission accomplished!
After a long descent, the maple seed
has landed far away.

Mission accomplished!
After a lively adventure, the chestnut seed
has also rolled far from where it fell.

So this maple seed grows up
under a different kind of tree.

And that chestnut seed sprouts
under a different kind of tree.

Other plants that fall . . .

and bounce:

Linden

and whirl . . .

Ash

Oak

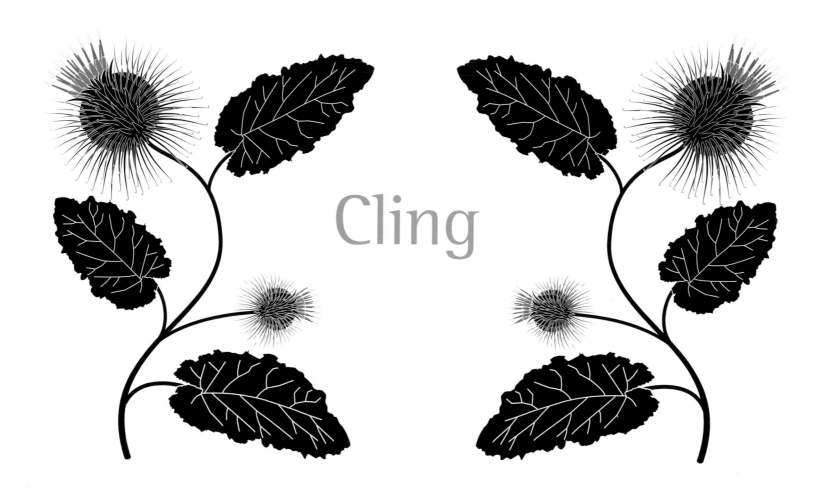

Cling

I'm a great burdock.
Insects love my pretty flowers,

which soon become little fruits:
spheres covered with tiny hooks.
What do they do?

They cling to things, of course!
An animal just has to walk by,
and my seeds get on board!
Where will the fox end up . . . ?

But all of a sudden . . .
Oops—the burr is spotted
by the fox! Too itchy!

Bye, fox!

The trip stops here. . . .

Well, almost!

Other plants that cling:

Catchweed
Bedstraw

Avens

Wall Barley

Get
Eaten

We're elderberry and blackberry, two very different types of bushes.

I'm a blackberry bush. I spread my long, thorny branches into the ground or into hedges.

I'm an elderberry bush. My branches reach upward— I'm almost a tree.

We may be different in some ways, but we both produce tasty fruit for animals to eat!

My berries attract birds, and the birds swallow them eagerly. Each elderberry has three little seeds hidden inside.

My berries attract birds and rodents. Blackberries are made of drupes: little plump balls that each contain a single seed.

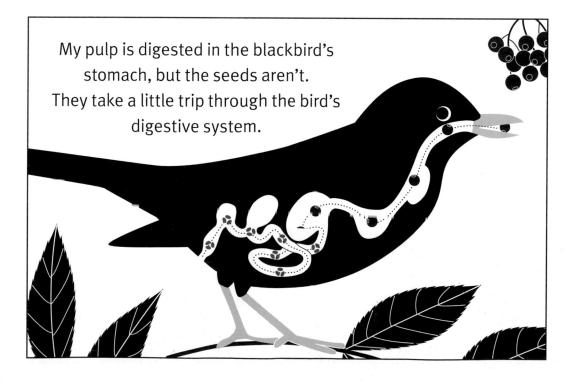

My pulp is digested in the blackbird's stomach, but the seeds aren't. They take a little trip through the bird's digestive system.

A similar trip happens when a field mouse eats a blackberry.

The blackbird
flies around,
and then . . .

The field mouse
goes home,
but before that . . .

The elderberry seeds get energy from their nutrient-rich cocoon so they can sprout.

The tiny blackberry seeds do the same.

Thanks to our technique of producing tasty fruit . . .

that animals love to eat, our seeds were able to travel!

Other plants with parts that are eaten:

Grapevine

Wild Service Tree

Ivy

And other plants with fleshy fruit
like Mistletoe, Cherry Tree,
Plum Tree, Apricot Tree, Sorb Tree,
Tomato, Apple Tree

Explode!

I'm a violet. I want to be everywhere, so I cover as much of the ground as I can!
For that to happen, I have my own special method.

Here's my flower.

It transforms into a fruit capsule
that holds my seeds.

When the fruit ripens, it opens
into three sections.

As the sections dry out, my seeds get a bit cramped.
Then . . . 5, 4, 3, 2, 1, blastoff!

What a launch!
Soon afterward, small violets
begin to grow all over.

Other plants that explode:

Wood
Sorrel

Hairy
Bittercress

Touch-Me-Not

Planted
by Animals

We're hazel and greater celandine.
We don't look anything like each other.
One of us is a shrub or tree, and the other
is a small, flowering plant.

But once our seeds are ripe, the way we travel
is similar. Animals take us on a journey.

I'm hazel, and
I produce many hazelnuts.
You don't need to pick them.
They fall off all by themselves.
Squirrels love them!

I'm greater celandine,
and each of my seeds has
a little white part that is
good to eat. Once the seeds
are ripe, they fall at my feet.

Ants can't resist them!
They carry them away. . . .

Hazelnuts keep for a really long time.
Squirrels hide lots of hazelnuts underground when
they're gathering their winter stash.

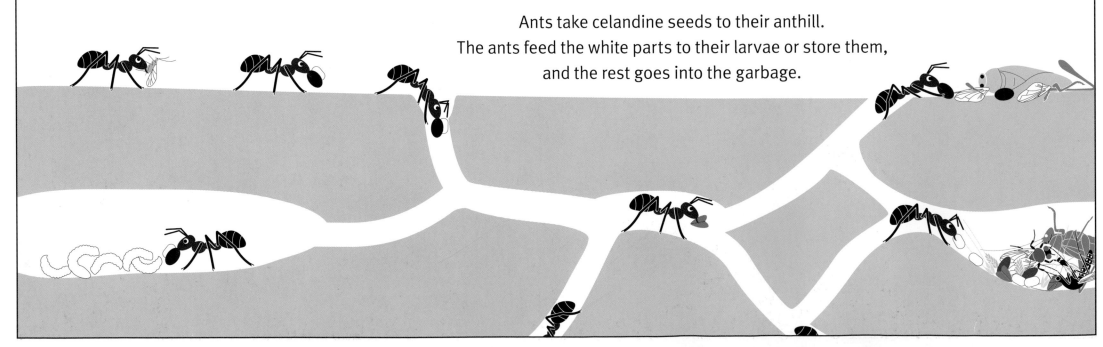

Ants take celandine seeds to their anthill.
The ants feed the white parts to their larvae or store them,
and the rest goes into the garbage.

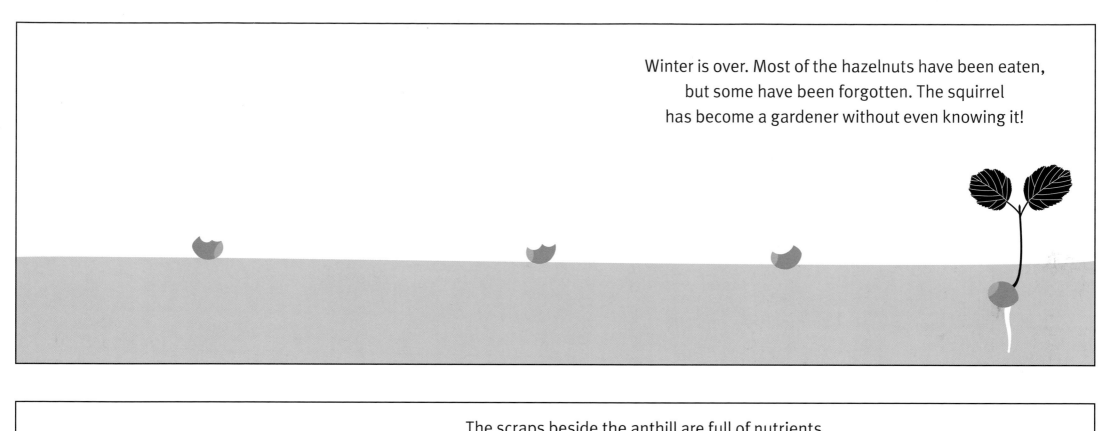

Winter is over. Most of the hazelnuts have been eaten,
but some have been forgotten. The squirrel
has become a gardener without even knowing it!

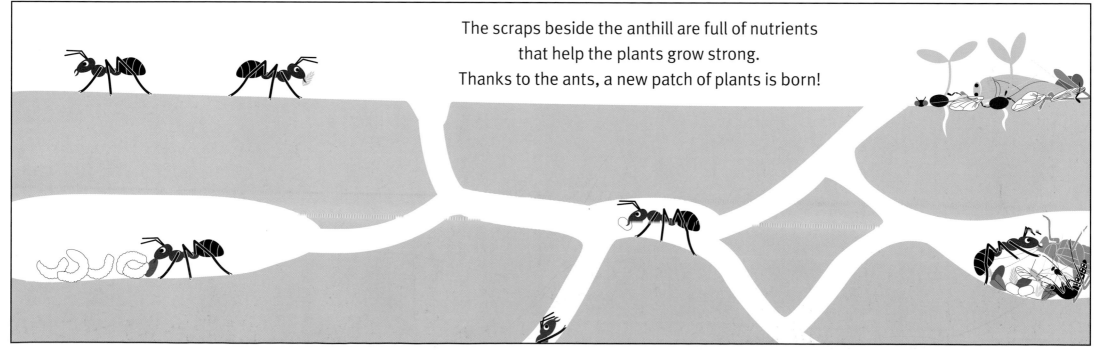

The scraps beside the anthill are full of nutrients
that help the plants grow strong.
Thanks to the ants, a new patch of plants is born!

These are also planted . . .

by squirrels

Walnut Tree

Beech Tree

and by ants:

Snowdrop

Wood Anemone

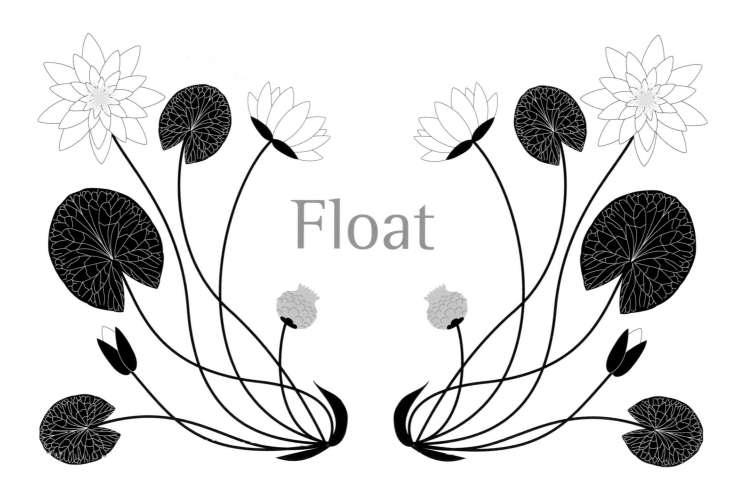

Float

I'm a white water lily. My roots are in the soil. My flowers and leaves are born underwater, but they continue to grow up at the surface.

After my flower is pollinated, it turns into a round, plump fruit.

Once it's ripe, the fruit floats, and sails away. . . .

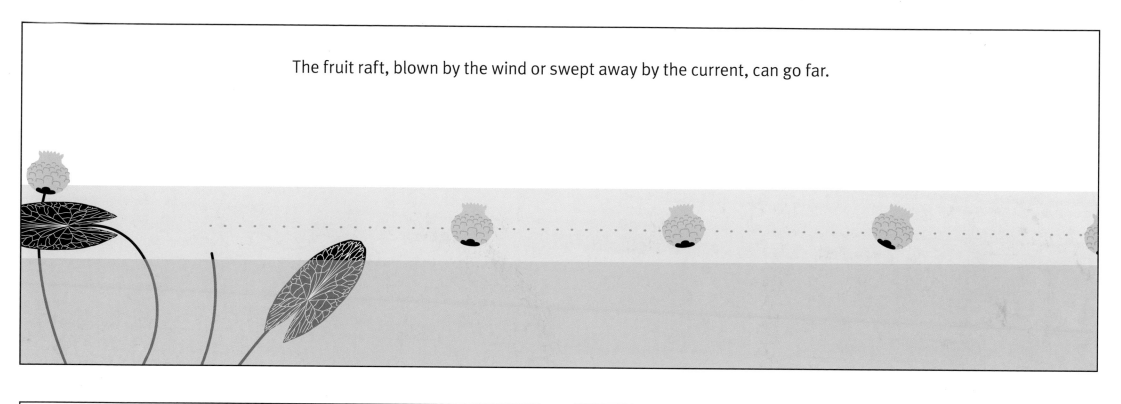

The fruit raft, blown by the wind or swept away by the current, can go far.

After some time in the water, it decomposes and frees all of the seeds that it was carrying.

And then this happens.

Other plants that float:

Arrowhead

Bur Reed

Yellow Iris

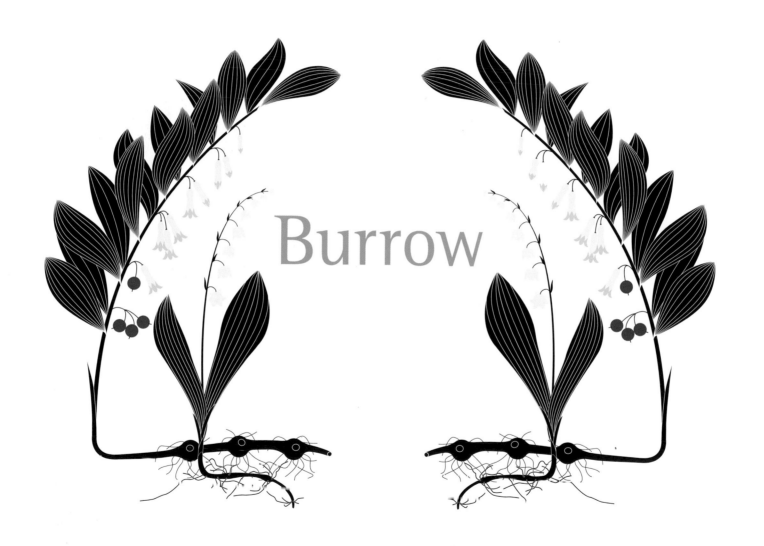

Burrow

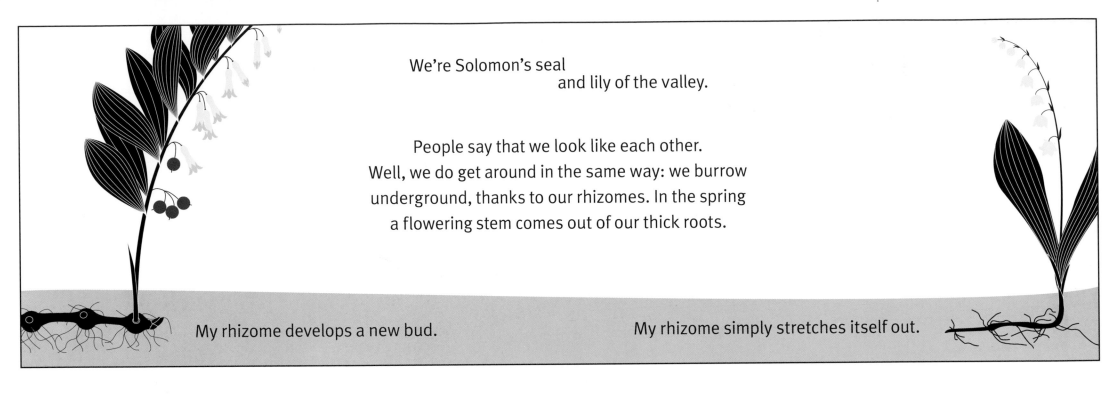

We're Solomon's seal
and lily of the valley.

People say that we look like each other.
Well, we do get around in the same way: we burrow
underground, thanks to our rhizomes. In the spring
a flowering stem comes out of our thick roots.

My rhizome develops a new bud.

My rhizome simply stretches itself out.

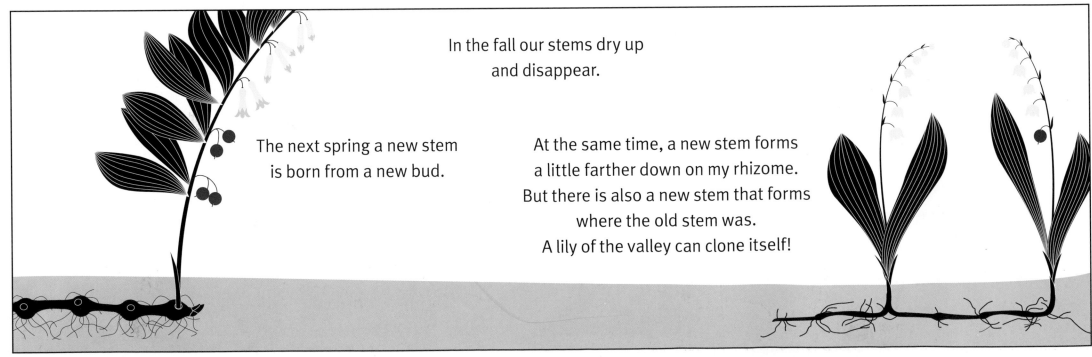

In the fall our stems dry up
and disappear.

The next spring a new stem
is born from a new bud.

At the same time, a new stem forms
a little farther down on my rhizome.
But there is also a new stem that forms
where the old stem was.
A lily of the valley can clone itself!

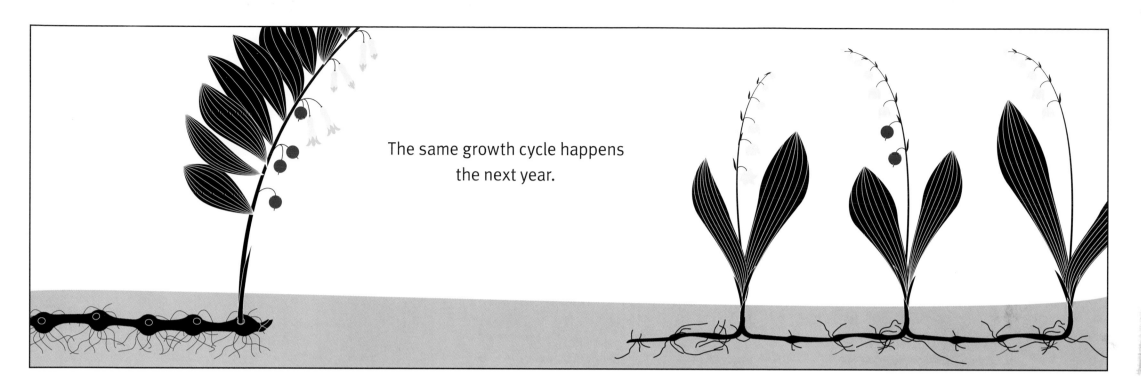

The same growth cycle happens
the next year.

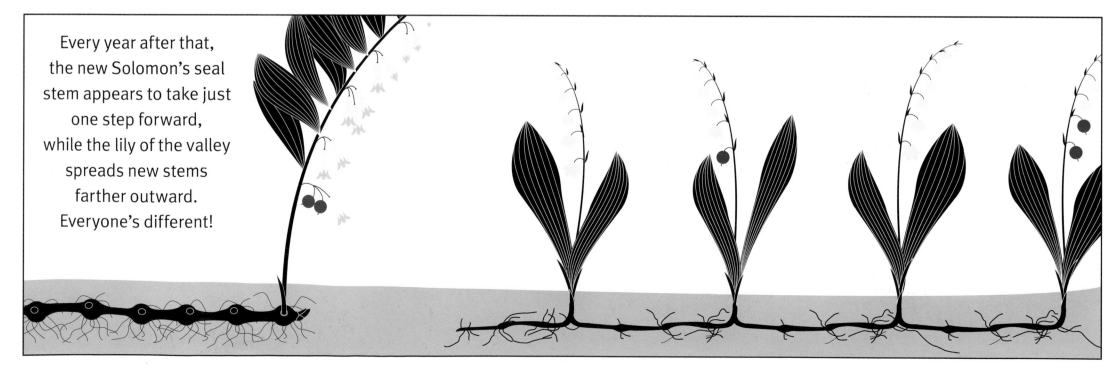

Every year after that,
the new Solomon's seal
stem appears to take just
one step forward,
while the lily of the valley
spreads new stems
farther outward.
Everyone's different!

Other plants that burrow:

Paris
Quadrifolia

Clover

Horsetail

Cultivated

I'm a bean.
Humans help me get where
I need to go. For example,
Camille's mom plants me
in her garden.

She gives Camille some of my seeds
to bring to school.

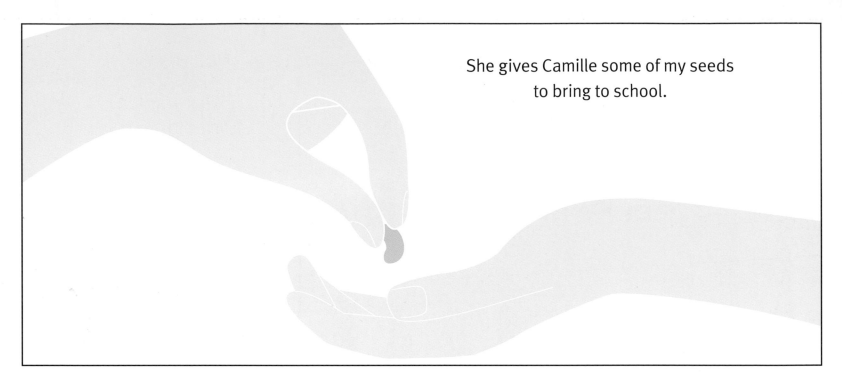

In class, children plant them
in little jars.

Then they bring
them home and put them
in window boxes . . .

or in their gardens.

What a big trip
for a little bean!

And that's really nothing when you consider
that long ago I arrived here in this very same way
from Central and South America!
Now you can find me everywhere. Humans have
helped me travel all over the world—by boat,
by truck, by plane, or by car!

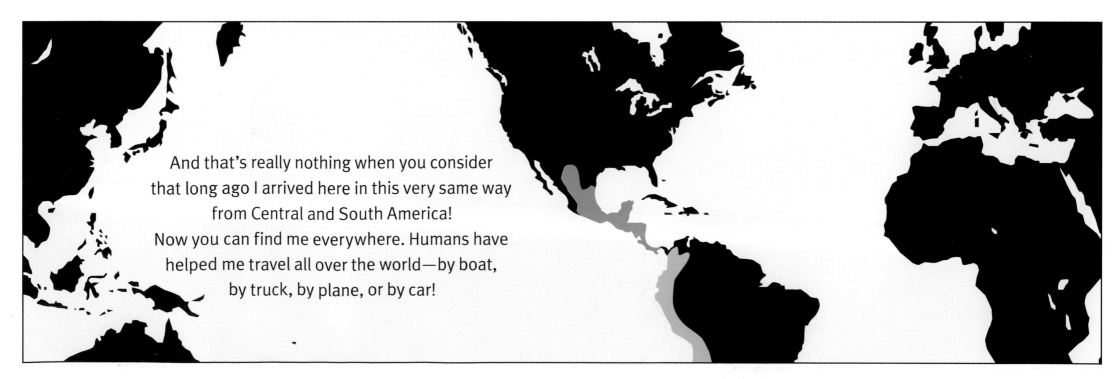

Humans have helped many plants travel around the world—
either to be used as food or to make their gardens look beautiful.
Here are a few examples, with their areas of origin:

Europe	Middle East	Asia	South and Central America
Columbine	Wheat	Cucumber	Begonia
Beet	Carrot	Kiwi	Squash
Black Currant	Onion	Chestnut	Corn
Cherry	Olive	Pear	Marigold
Cabbage	Pea	Apple	Pepper
Raspberry	Tulip	Rice	Tomato

Some plants use more than one method to scatter their seeds. The violet produces stolons, some of its seeds explode outward, other seeds ripen at its feet, and yet others are carried away by ants. The white poplar's seeds fly away and float on water. The strawberry plant makes stolons, and its seeds are also eaten!

Here are the scientific names of all the types of trips that plants take:

- Plants that fly: anemochory (by the wind)

- Plants that are cultivated: anthropochory or hemerochory (by humans)

- Plants that explode, burrow, or creep: autochory (by the plants themselves)

- Plants that fall: barochory (by gravity)

- Plants that float: hydrochory (by water)

- Plants that are planted by animals, plants that cling, plants that are eaten: zoochory (myrmecochory, by ants; ornithochory, by birds; and mammalochory, by mammals)

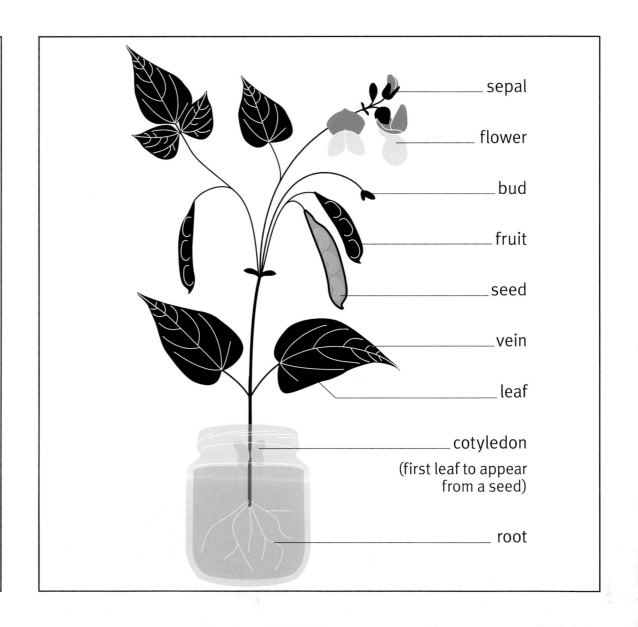

sepal

flower

bud

fruit

seed

vein

leaf

cotyledon
(first leaf to appear
from a seed)

root